PRETTY GOOD SCIENCE JOKES

**230 Peer-Reviewed,
Lab-Tested, Nobel-Worthy
Jokes, Puns, and Zingers**

Edited by Steve Mockus
Illustrations by Johnny Sampson

CHRONICLE BOOKS
SAN FRANCISCO

For Dad and for Mom, and for Jody.
—S.M.

For my parents, who taught me what a groaner was,
and would still laugh whenever I told one.
—J.S.

Library of Congress Cataloging-in-Publication Data available.

ISBN 978-1-7972-3728-2

Manufactured in China.

Illustrations by Johnny Sampson.
Design by Barbara Bersche.

10 9 8 7 6 5 4 3 2 1

Chronicle books and gifts are available at special quantity discounts to
corporations, professional associations, literacy programs, and other organizations.
For details and discount information, please contact our premiums department at
corporategifts@chroniclebooks.com or at 1-800-759-0190.

Chronicle Books LLC
680 Second Street
San Francisco, CA 94107
www.chroniclebooks.com

CONTENTS

PHYSICS

Did you lose an electron?

You really have to keep an **ion** them.

What do you call a scientist who studies carbonation in soft drinks?

A **fizzicist**.

Entropy:

It isn't what it used to be.

What does a subatomic duck say?

"Quark."

A group of protestors walks into a physics lab and chants:

What do we want?
Time travel!
When do we want it?
Irrelevant!

What did one magnet say to the other?

"I find you very **attractive**."

I'm not lazy—
I'm in a state of **potential** energy.

My new sweater kept picking up
static electricity, so I took it back
to the store.

They gave me another one,
free of charge.

Two atoms are walking down the
street and one says to the other,
"Help, help, I've lost an electron!"

The second atom says,
"Are you sure?"

And the first atom says,
"Yes! I'm **positive**!"

Where do bad rainbows go?

To **prism**.
It's a **light** sentence,
but it gives them time to **reflect**.

Heisenberg was cruising in his car and definitely speeding when he got pulled over by a police officer. The officer was very angry and came up to Heisenberg and said, "Do you have any idea how fast you were going?" And Heisenberg said, "No, but I know **where** I am."

What sort of convertible bed does the smallest packet of electromagnetic energy sleep on?

A **photon**.

What do you call the arrival of the first swine in the universe?

The **Pig** Bang.

What did the proton say to the electron to start a fight?

"I'm sick of your **negativity**."

What do electricians have for breakfast?

Ohmlettes.

Why did the man with a headache turn on a bunch of radios?

Because they **kil-o-hertz**.

How do you count atoms?

You **atom** up.

What song does an electric
cowboy sing?

"Ohm on the Range."

A desperate student got caught
cheating on his physics exam,
and he's in so much trouble now.

He didn't understand the **gravity**
of the situation.

Have you read that new book
about antigravity?

It's impossible to put **down**!

Friction: It's such a **drag.**

How do nuclear scientists relax?

They go **fission**.

A photon checks into a hotel.
The bellhop asks if they can help
with the luggage. The photon replies,
"I don't have any. I'm **traveling light**."

What keeps a dock floating
on the water?

Pier **pressure.**

What's a physicist's favorite food?

Fission chips.

Why was the electrician
a great conversationalist?

She was always up on **current** events.

Never trust atoms.

They **make up everything**.

A proton, neutron, and electron walk into a café. The proton orders a drink, takes out its wallet, and pays.

The electron orders a drink, takes out its credit card, and pays.

The neutron orders a drink and is about to pay when the bartista says, "Wait! **No charge**."

Why did the scientist take out his doorbell?

Because he wanted to win the **no bell prize**.

What travels faster, heat or cold?

Heat, because you can
easily **catch cold**.

The real facts about electricity
might **shock** you.

What do you call a scientist
who steals energy?

A **joule** thief.

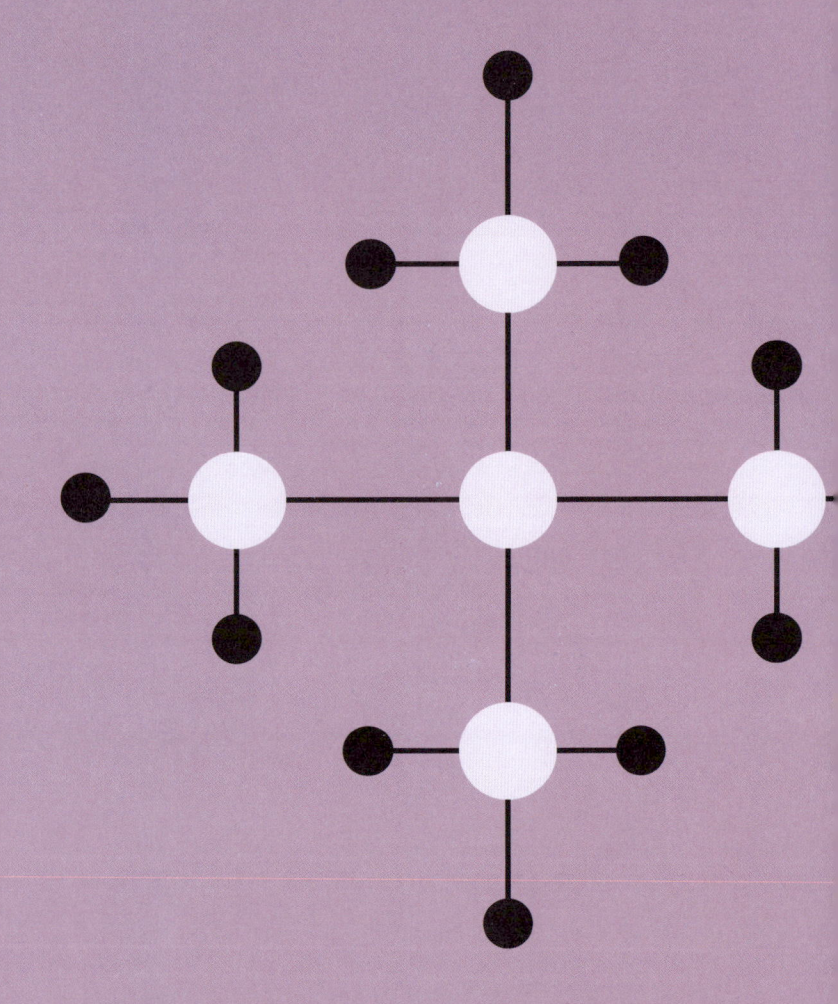

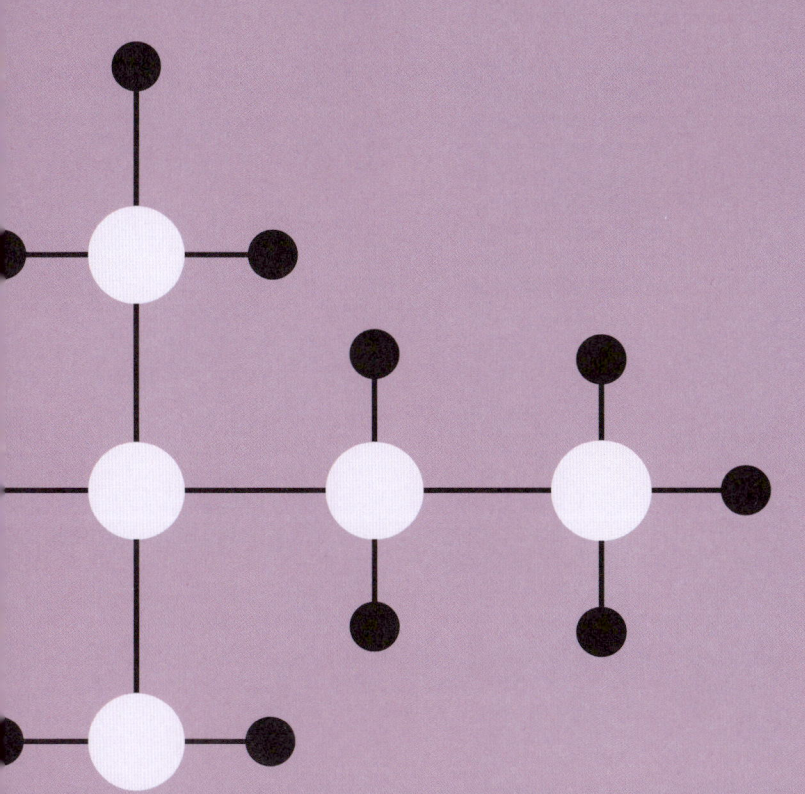

CHEMISTRY

I tried to find a good chemistry joke, but all the good ones **argon**.

I'm told my chemistry jokes are like cobalt, radon, and yttrium.

CoRnY.

What did the chemist say when he heard oxygen and magnesium were dating?

"OMg."

Oxygen and potassium went on a date.

It went **OK**.

Why are chemists great at solving problems?

Because they have all the **solutions**.

How do you make antifreeze?

Hide her blanket.

What kind of dogs do chemists prefer?

Laboratory retrievers.

Why did the hipster chemist burn his hand?

He touched the beaker before it was **cool**.

Want to hear a joke about nitrogen oxide?

NO!

What did the scientist say to the chemist whose lab smelled like rotten eggs?

"Sorry for your **sulfering**."

What do you say to boiling water?
"You will be **mist**."

Anybody know a good joke about sodium?

Na . . .

Do you need any sodium hypobromite?

NaBrO.

What type of fish is made out of two
sodium atoms?

2 Na.

What do you get after
sixteen sodium atoms?

NaNaNaNaNaNaNaNaNaNaNaNaNaNaNaNa
BATMAN.

Are you made of copper and tellurium?

Because you are a **Cu Te**.

Did you know Neil Diamond
was called Neil Coal before
the **pressure** got to him?

What's a pirate's favorite element?

Aaaaargon.

Do you have eleven protons?
Because you are **sodium cute**.

What do you call a ring of iron atoms
at a carnival?

A **ferrous** wheel.

Where did the chemist eat his lunch?

On a **periodic table**.

What is a chemist's favorite holiday song?

"O **Chemist** Tree, o **Chemist** Tree."

Are you made of sulfur, lanthanum, and yttrium?

Because you **SLaY** all day.

What do you do with a dead chemist?

You **barium**.
But only if you can't
helium or **curium**.

I told a chemistry joke.

But there was **no reaction**.

H_2O is water, but what is H_2O_4?

**For drinking, washing,
and swimming in**.

Why do chemists prefer nitrates?

They are cheaper than **day rates**.

What do you call your favorite piece
of equipment in a lab?

The **best** tube.

If you swallow uranium,
would you get **atomic ache**?

Why did the two hydrogen molecules
break up?

They had a **weak attraction**.

I'm out of chemistry jokes.
I should **zinc** of a new one.

Ionestly can't think of any more
chemistry puns.

BIOLOGY

Whenever I'm around you, I undergo anaerobic respiration.

Because you **take my breath away.**

I may look like I'm doing nothing, but I am very busy on the **cellular level**.

Why did the bacteria cross the microscope?

To get to the other **slide**.

What do you call it when a cell takes a picture of itself?

A **cell**fie.

What do zebras have that no other animals have?

Baby zebras.

Why do storks raise one leg?

Because if they raised **both** legs, they'd fall down.

How did Pavlov get his hair so soft?

He **conditioned** it.

How do trees surf on the internet?

They **log** in.

How do you make a tree laugh?

Tell it **acorn**y joke.

How do plants make
themselves heard?

With amp**leaf**ication.

Would you like a quick explanation
of an acorn?

In a **nutshell**, it's an oak tree.

What's the difference between a dog and a marine biologist?

One **wags a tail**;
the other **tags a whale**.

How many evolutionary biologists does it take to change a light bulb?

Just one, but it takes
millions of years.

Why did the biologist break up
with the physicist?

They had **no chemistry**.

What did the microbiology student
get for being late to class?

A **tardigrade**.

What did the femur say to the patella?

"I **kneed** you."

Why don't ants get sick?

Because they have little **anty bodies**.

Why is the nervous system reckless?

It does everything on **impulse**.

Anatomy jokes: so **cornea**, but also **humerus**.

I would give you flowers.
But I never **botany**.

What kind of plant grows in
your hand?

A **palm** tree.

When a plant is sad, do other plants **photosympathize**?

Why was the mushroom so popular?

He was a **fungi**.

Amateur mycologists have
questionable morels.

How much room do fungi need
in order to grow?

As **mushroom** as possible.

What do genetically modified
horses eat?

DNhay.

What's the best way to prevent disease from biting insects?

Don't **bite** any.

What did the procrastinating zoologist say about inspecting the reptile house?

"**Iguana** do it tomorrow."

Why don't you ever see elephants
hiding in trees?

Because they are so good at it.

How do corvids stay together
in flight?

With vel**crow**.

How many tickles does it take an octopus to laugh?

Tentacles.

Why are fish so good at science?

They spend all their time in **schools**.

Where do killer whales go
for braces?

To the **orca**dontist.

Where did the hippo go to college?

At the **hippocampus**.

What do plants eat between meals?

Light snacks.

Why did the tree get in trouble?

It was being **knotty**.

What's the best vehicle for leaf watching in the fall?

An **autumn**mobile.

Why do seagulls fly over the sea?

Because if they flew over the bay, they would be called **bagels**.

How does a good marine biologist work?

With **a-fish-in-sea**.

Why did the manta ray want to talk
to the diver?

He wanted to have a
manta-man conversation.

What did the tree wear to his friend's
pool party?

Swimming **trunks**.

It's been scientifically proven that people who celebrate more birthdays **live longer**.

What do you call the gradual development of organisms at the bottom of the ocean?

Evo**low**tion.

Why are herpetologists the most inquisitive zoologists?

Because they **axolotl** questions.

Why do birds fly south for
the winter?

It's too far to **walk**.

What do you call a hibernating
creature caught in a spring shower?

A **drizzly bear**.

How do bears stay cool
in the summer months?

They use **bear conditioning**.

What's the difference between a
panda bear and a polar bear?

About **4,000 miles**.

Why was the amoeba sad?

His parents just **split**.

Why do biologists look forward to casual Fridays?

They can wear their **genes** to work.

What did one cell tell his sister cell
when she stepped on his toe?

"Ouch! That's **mitosis**."

These biology puns are so good they
cell themselves.

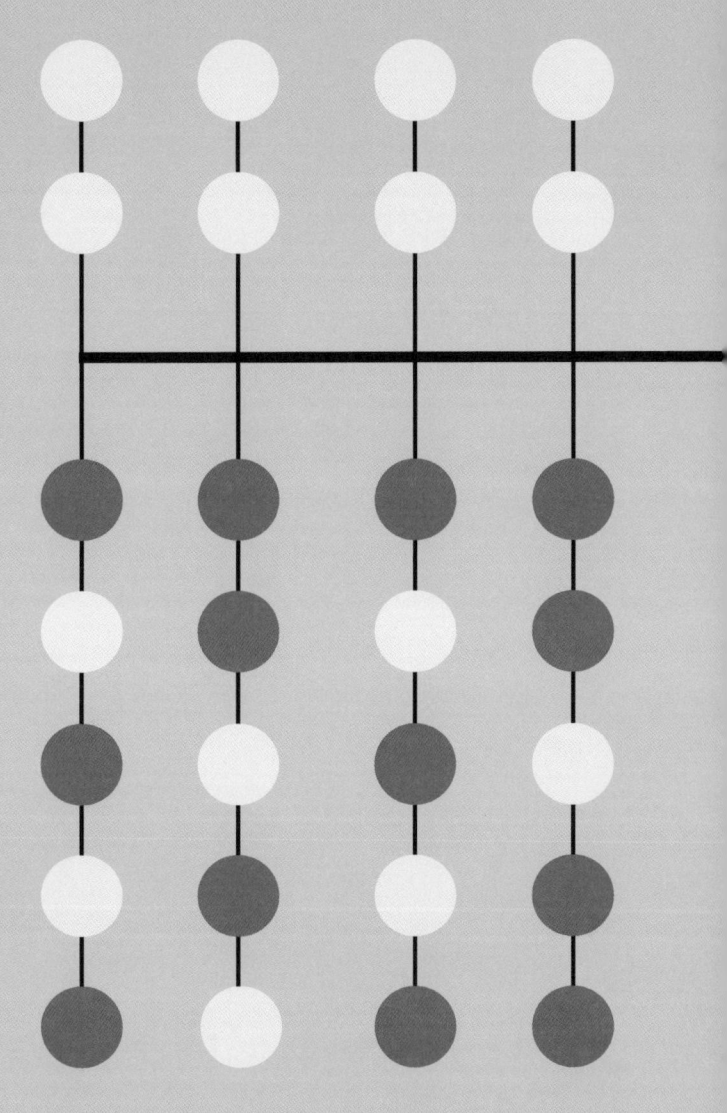

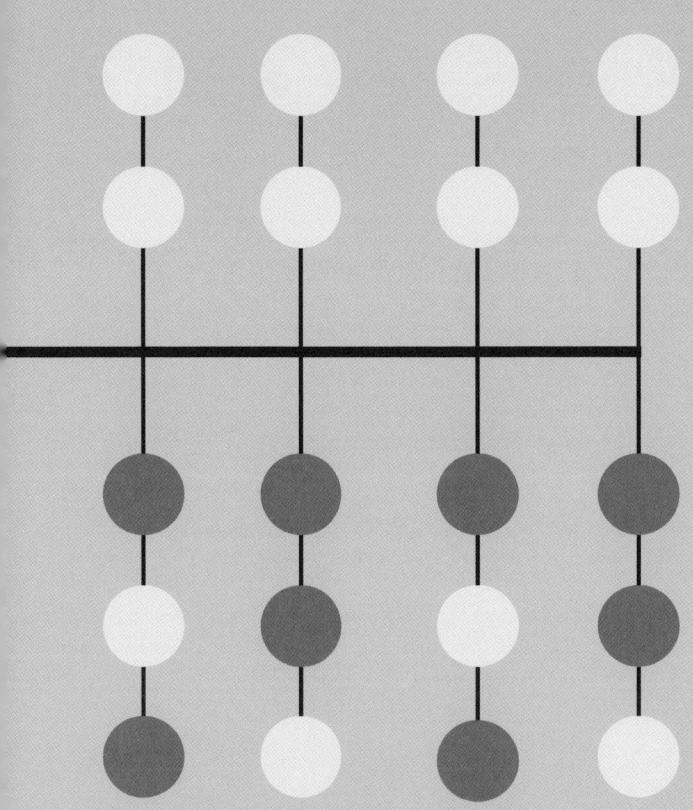

MATH &
COMPUTER SCIENCE

Not all math puns are terrible.

Just **sum**.

The best angle from which to approach any problem?

The **try**angle.

What works faster than a calculator?

A calcu**now**.

What kind of beats do
mathematicians like to dance to?

Loga**rithms**.

Why does the obtuse triangle never win arguments?

Because it's never **right**.

What is an algebra teacher's favorite sandwich?

A **slope-y joe**.

Why did the obtuse triangle go
to the beach?

It was over **90 degrees**.

Who adds, subtracts, multiplies, and bumps into light bulbs?

A **moth**ematician.

What kind of math do owls like best?

Owlgebra.

What type of math is discussed
between seagulls?

Intergull calculus.

What do you call a hen who counts
her own eggs before they are
hatched?

A mathema**chicken**.

Why was 6 afraid of 7?

Because **7 8 9**.

Do you know why 7 8 9?

Because you're supposed to eat **three squared** meals a day.

Why did the two 4s skip lunch?

They **already 8**.

I got in a fight with 3, 5, 7, 9, and 11.

The **odds** were against me.

To whoever invented the zero:
Thanks for **nothing**.

What do you call the longest side of a
right triangle in the forest?

A hypoten**moose**.

Why was the mathematician such a terrible gardener?

All his plants had **square roots**.

If puns make you feel numb,
do math puns make you feel **number**?

What do baby parabolas drink?

Quadratic **formula**.

What did one math book say
to the other?

"Leave me alone.
I have **my own problems**."

Why are bacteria bad at math?

They **multiply by dividing**.

Where do mathematicians sit
at the banquet?

At the **multiplication** table.

Why couldn't the mathematician
afford lunch?

He could **binomials**.

Have you heard the joke
about the statistician?

Probably.

What's a mathematician's
favorite season?

Summer.

What do you call an empty
parrot cage?

A **polygon**.

What did the bee say when it solved
the problem?

"**Hive** got it!"

Are monsters good at math?

They are if you **count Dracula**.

What is the best tool to use
for math?

Multi**pliers**.

How does a mathematician
plow fields?

With a pro**tractor**.

Why did the geometry teacher
miss class?

Because she **sprained her angle**.

———
———

Parallel lines have so much
in common.

It's a shame they'll **never meet**.

Why should you never start a conversation with pi?

Because it'll **never end**.

What do you get if you divide the circumference of a jack-o'-lantern by its diameter?

Pumpkin **pi**.

Did you hear about the mathematician who is afraid of negative numbers?

She'll **stop at nothing** to avoid them.

Hardware: the part of the computer that you can **kick**.

What do computer coders do when they stub their toes?

HTM**yell**!

There are 10 kinds of people in this world: those who **understand binary** and those who don't.

What kind of beats do computer programmers like to dance to?

Algo**rithms**.

Why do programmers prefer to code in the dark?

Because light attracts **bugs**.

What diet did the ghost developer go on?

Boolean.

Why was the spider on
the computer?

It was updating its **web**site.

How do computers learn?

Bit by bit.

Why did the programmer quit
their job?

Because they didn't get **arrays**.

Users are advised not to use *beefstew* as their password.

It's not **stroganoff**.

I don't like computer science jokes.

Not one **bit**.

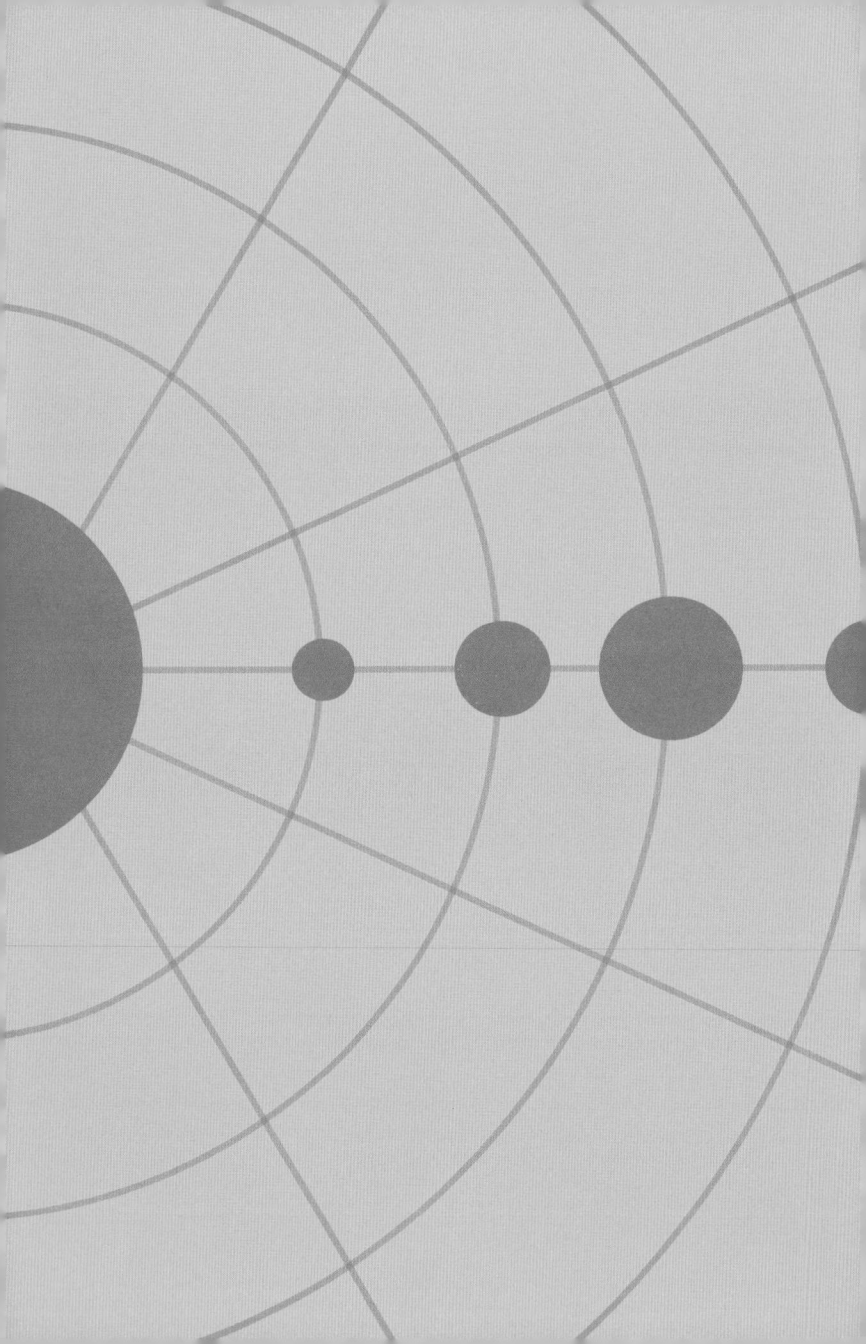

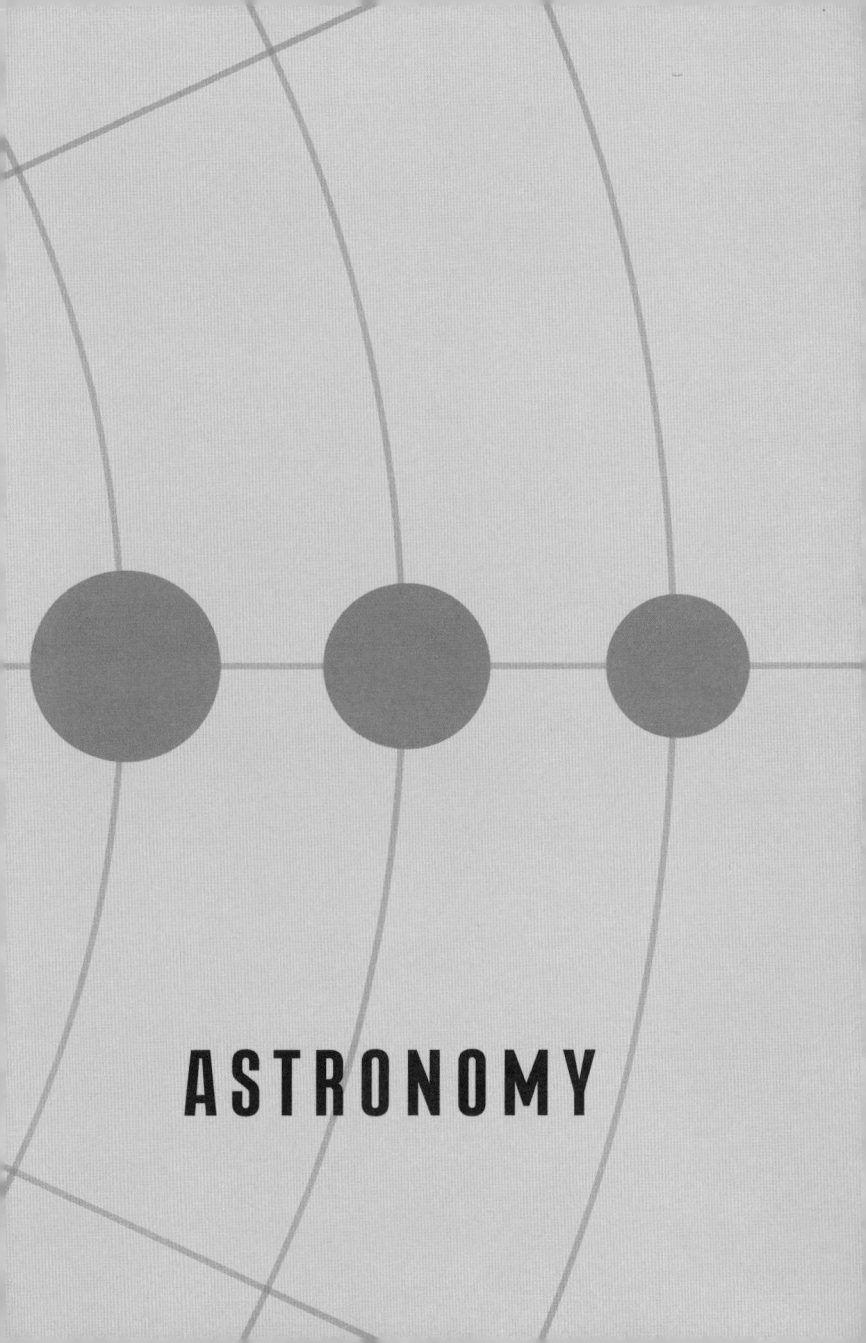

ASTRONOMY

Space was cool before it **mattered**.

Why did the student fail the astronomy class?

It was all **over his head**.

How do you organize a party
for astronomers?

You **planet**.

Why do astronomers take a positive
view of everything?

They're all **starry-eyed**.

What do you call an imploded star
between 34th and 35th Streets?

A **block** hole.

Why aren't there any cats in space?

Because cats abhor **vacuums**.

Where do astronauts park their ship?

At a parking **meteor**.

What did the asteroid say when the reporter asked him a question?

"No **comet**."

How does our solar system hold up its pants?

With a **meteor belt**.

What kinds of books do planets
usually like to read?

Comet books.

What kind of music do planets
dance to?

Nep**tunes**.

What's the best way to finance a trip to the moon?

Buy now, pay **crater**.

I love astronomy! The rotation of the Earth really **makes my day**.

Early astronomers couldn't figure out where the sun went.

But then it **dawned** on them.

What did the therapist say
to the moon?

"Don't worry, I think you're just
going through a **phase**."

Why wasn't the moon hungry?

Because it was **full**.

Did you hear about the restaurant
on the moon?

The food is out of this world, but
there's **no atmosphere**.

What is the sun's other job?

Moonlighting.

What would happen if the universe ceased to exist?

No **matter**.

Where do rocket chips travel to?

The **Big Dipper**.

Why didn't the Dog Star laugh at the space joke?

It was too **Sirius**.

How do you illuminate space?

With a satel**lite**.

Why didn't the sun go to college?

Because it already had
three million **degrees**.

What is the center of gravity?

The letter **V**.

In 1905, Einstein developed
a theory about space,
and it was **about time** too!

What is an astronomer's
favorite food?

Mercurry.

What's an astronaut's
favorite beverage?

Gravi**tea**.

I was charged $5,000 for sending my cat into space. It was a **cat astro fee**.

There is no precipitation on the moon, except when it's **waning**.

What do you call a lazy cosmic explorer?

A **procrastronaut**.

When do astronauts eat?

At **launch** time.

Why didn't the astronomer get her work done?

She just couldn't get into the **ozone**.

Why haven't aliens visited Earth?

It's only rated **one star**.

Living on Earth may be expensive, but it does include a free **trip around the sun** every year.

I have pressed the **space bar** fifteen times now, but I'm still here stuck sitting at my desk.

Knock knock.
Who's there?
Solar.
Solar who?

Solar you going to think of a better astronomy joke?

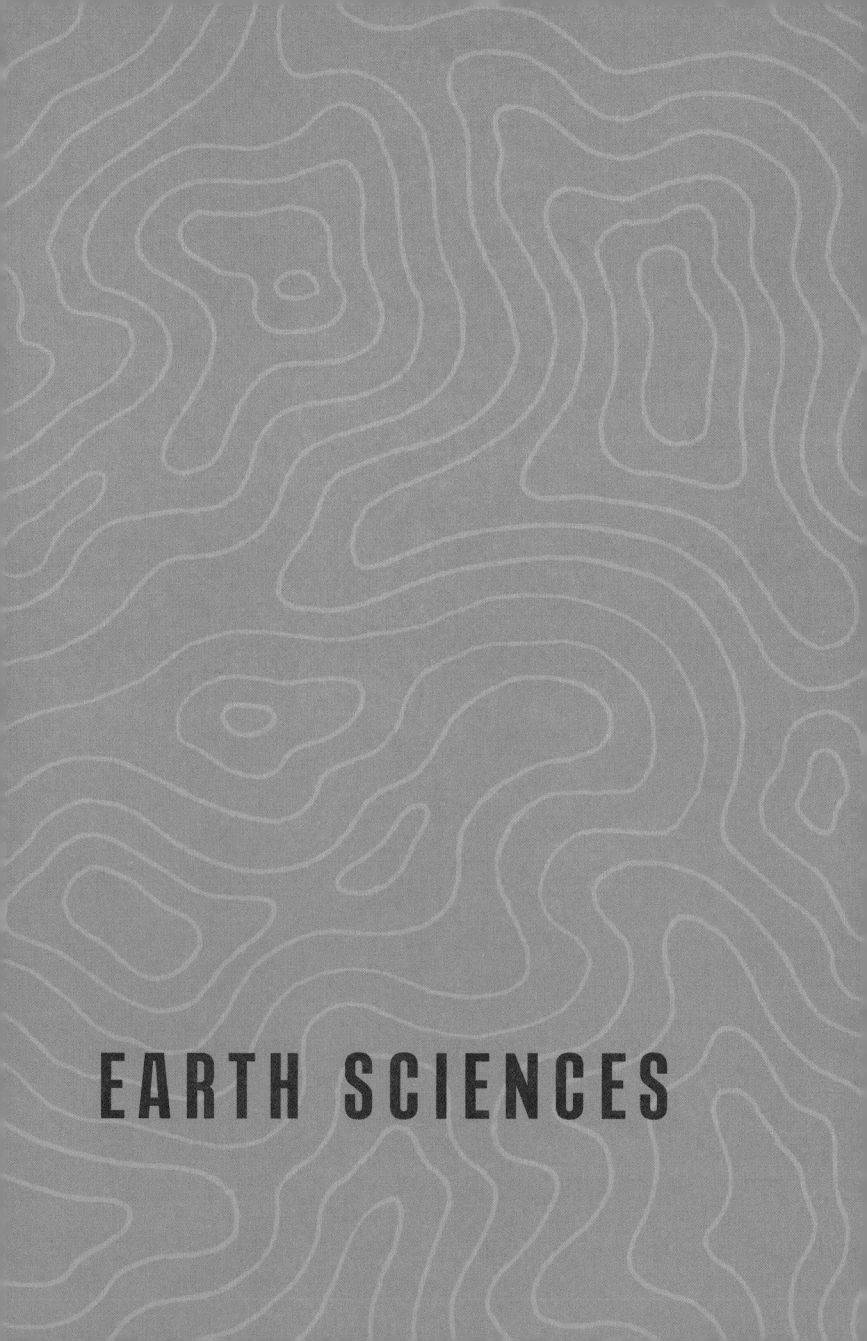

EARTH SCIENCES

Geology puns are great. They really bring **pebble** together.

Why couldn't the tectonic plates maintain a relationship?

There was too much **friction** between them.

There's no halfway with a geologist.

It's all **ore** nothing.

Why couldn't the geologist think
of a joke?

It was on the tip of her **tungsten**.

Where do geologists go to relax?

Rock concerts.

When are geologists unhappy?

When people take them for **granite**.

Don't expect perfection
from geologists.

They all have their **faults**.

How do geologists
ask each other out?

"Are you a carbon sample?
Because I'd love to **date** you."

Where do minerals like to sleep?

In **bed**rocks.

Why was the sedimentary
rock cheap?

Because it was on **shale**.

Geology rocks, but geography is
where it's really at.

Why was the student encouraged to study oceanography?

Because his grades were all below **C level**.

What did the ocean say to the river?

"You can run, but you can't **tide**."

Why is the ocean so salty?

The land never **waves back**.

These geology jokes work on so many **layers**.

What did one tectonic plate say when he bumped into the other?

"Sorry! **My fault**."

Why was the gemstone afraid
of taking his exam?

Because he thought he wasn't
going **topaz**.

Quartz the matter? Can't handle
these rock puns?

Anyone want to buy a broken barometer?

No **pressure**.

What did one raindrop say to the other?

"Two's company. Three's a **cloud**."

What color is the wind?

Blew.

Why did the cloud date the fog?

Because it was so **down to earth**.

What is the difference between weather and climate?

You can't **weather** a tree, but you can **climate**.

Why can't you play hide-and-seek
with a mountain?

Because the mountain **peaks**.

Why was the young cloud always
in trouble?

Because it never took
anything **cirrus**ly.

Why are oceanographers always
on time?

They like to stay **current**.

Why do corals get stressed?

Current events.

Where do saltwater fish look
for jobs?

In the **kelp-wanted** ads.

Why wouldn't the sea creatures
share their treasure?

They were **two shellfish**.

Why did the dolphin cross
the ocean?

To get to the other **tide**.

What game show do
oceanographers like best?

Whale of Fortune.

ACKNOWLEDGMENTS

STEVE MOCKUS is a San Francisco–based editor and author of the *Classic Horror Oracle, Cthulhu: The Ancient One Tribute Box, Mr. Spock Logic & Prosperity Box, How to Speak Zombie, Eerie Legends, A Pocket Dictionary of the Vulgar Tongue,* and *Stick Man's Really Bad Day.*

JOHNNY SAMPSON is an award-winning cartoonist and illustrator based in Glen Ellyn, Illinois. He has been the sole writer and artist of the *MAD* Fold-In since Al Jaffee retired in 2020. When he's not hunched over a drawing table, he enjoys playing piano and guitar and carving around in bowls on his skateboard, though not all at the same time. See more of his work at www.johnnysampson.com.

THANK YOU
FOR YOUR SCIENCE JOKES

Malcolm Billingsley
Juliette Capra
Kenan Chan
Perry Crowe
Ryan Cunningham
Jason Hammond
Luke Hammond
Mirabelle Korn
Jody Mosley
Megan Reasor
Baron Roberts
Aaron Sutliff
Jackson Twilling